I Wanna Go to Grandma's House

Written by
Grandma Janet Mary Sinke

Illustrated by
Craig Pennington

My Grandma and Me *Publishers*

Acknowledgements

Special Thanks to:

Lynne Koenigsknecht, of Radiant Designs, for her support and many hours of work designing the web page.

Darlene Koenigsknecht, for all her dedicated hours of work organizing the many aspects of this project.

Carmela D'Alessandro and Lorraine Kudwa for editing.

Sandra Steffen, published author, for her support and suggestions, for she believed from the start in the magic of this story.

––––––––––

Grandma Janet Mary ™
My Grandma and Me Publishers ™
P.O. Box 144
St. Johns, Michigan 48879
FAX: 989-224-3749
Web Site: www.mygrandmaandme.com
E-Mail: info@mygrandmaandme.com

First Edition
Second Printing, April 2004
Printed and bound in Canada
Friesens of Altona, Manitoba

Library of Congress Cataloging-In-Publication Data on File

ISBN 0-9742732-0-1
LCCN 2003107647

Author's Dedication
To . . .

. . . my husband, Mike, whose love and support frees me to answer the Spirit.

. . . my children, the pride, the joys of my life.
Jason and Nicole
Emily and Kevin
Jake and Barb
Sarah and Mike
&
Michael and Kim

. . . my grandbabies, angels, who fill my heart with wonder,
Ellen Marie
Madalyn Mae
Ethan Jacob
&
all grandchildren yet to come.

—Grandma Janet Mary

Illustrator's Dedication
To . . .

. . . my wife Natalie, who inspires and challenges my creativity.
Her love and support encourage me to be the husband, father, and man
I strive to be.

. . . my sons, Owen and Guy, who fill my heart with love and joy.

. . . my parents, Jim and Kathie, who have been wonderful
role models as parents and extraordinary examples
of Godly grandparents.

—Craig Pennington

To grandchildren of all ages,

Visit your Grandma's house often.

Let her kiss your face.

Listen to her stories.

Make happy memories today.

That way, no matter how old you grow to be,

when your heart says,

"I Wanna Go to Grandma's House,"

your mind can take you there.

With Love,

Grandma Janet Mary

"I wanna go to Grandma's house,"
I told my mom today.

"Okay," she said,
"I'll call her up.
I know just what she'll say."

I listen on the other phone.
My Grandma's all excited.

❖

"Of course," she says,
"do bring the child.
What fun! I'd be delighted,
to have my little angel near.
She's pretty as can be,

❖

and I would love to have
her here to spend some time
with me."

Now,
all who know my Grandma say
she's sweet as cherry pie.
She calls me her dear angel girl
and here's her reason why.
"Your heart is kind and good," she says,
"and if the light is right,
yes!
I can see your halo shine
so beautiful and bright."

But,
I don't know if all that's true,
yet, one thing is for sure,
my Grandma thinks I'm pretty great
and I think that of her.

At last we're finally on our way
and I can hardly wait.

We turn the corner.
Grandma's there.

I see her. She looks great!

For there she is in her front yard,
she's wearing pink capris.
She says her legs look better if
she covers up her knees.

Grandma waves and claps her hands
as we pull in the drive.
She's very old, but still my friend.
I think she's fifty-five.

"My angel girl," she hugs me tight
and swings me all around.
I hug her back and let her take . . .

. . . my feet right off the ground.

Then Grandma gently puts me down.

She gives my face a kiss;

and when she hugs my mom,
I think . . .

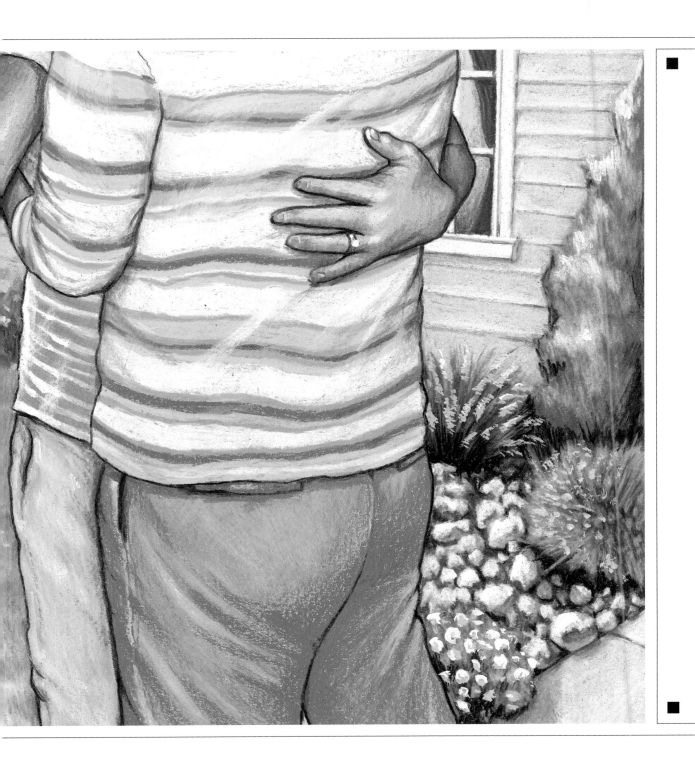

. . . it's nice to be like this.

"I'll see you hon'," my mom, she says,
"when I come back tonight."
Then I with Grandma wave good-bye
till she is out of sight.

Grandma smiles then says to me,
"Let's start this brand new day,
by baking something special in
my kitchen while we play."

We tie our yellow aprons on.
We wash up in the sink.
I get her baking spoon and bowl.
I'm ready now, I think,

❖

to mix the sugar with a clump
of softened yellow butter;
and all the while I work and stir
I hear my Grandma utter,

"You are the dearest angel girl,
that is my point of view.
Why, I just can't imagine what
without you I would do."

I think about what she just said
and yeah, I would agree.
'cause Grandma's
getting older now.
She needs someone like me.

I crack an egg. I make a mess.
But Grandma doesn't care.

She lets me crack another one,
then brushes back my hair.

It doesn't seem to matter
that the shells are in the dough.
"So what," she says,
"I'll never tell.
No one needs to know.

❖

And should your grandpa
take a bite
into some shells for lunch,
do not fret,
'cause cookies taste
much better with some crunch."

Grandma says,
"Let's take a break.
It's time to rock 'n roll."

❖

She's dancing
in the kitchen now.
WOW!
Grandma's got some soul!

I giggle 'cause it's funny when
she sings into a spoon.
She tells me it's a microphone
and then belts out a tune.

❖

At last the cookies are all done.
I ask, "What's there to do?"
I turn around.
She winks at me . . .

. . . her bathing suit is blue!

❖

"Well, hurry child;
get on your suit.
We'll go into the pool,
where we will swim
and
have some fun.
I'm glad you're not in school!"

Now,
Grandma doesn't look a lot
like girls who are real thin,
especially in her bathing suit.

❖

She's got some extra skin.

❖

But,
I think she looks beautiful.
She's Grandma after all.
Her extra skin is nice and soft.
I'm glad she's not too small.

The water's great.
I jump and dive.
I get my Grandma wet;

❖

and though
her face and hair are soaked,
she doesn't get upset.

We splash and play.
It's been such fun.
This morning's been the best.

❖

But Grandma says,
"It's time to eat,
and then we'll take a rest."

So,
after soup and sandwiches
we're underneath the tree.

❖

And from the hammock
we look up;
it's beautiful to see
skies of blue that peek on
through
the leaves of em'rald green

❖

and bluebirds curl up
in their nests
while we enjoy the scene.

I listen to my Grandma tell me
stories of back when,
my own dear mom
was growing up,
when she was nine and ten.

❖

And even though
these tales I've heard
so many times before,
I say to her all nestled in,

❖

"Dear Grandma, tell me more."
She smiles and talks of memories
that she will always keep,
while I just snuggle next to her
and then drift off to sleep.

I wake to find my Grandma
waiting patiently for me.
I slowly stretch.
She takes her time . . .

. . . to sip a cup of tea.

❖

The two of us then take a walk
out to the garden where . . .

. . . she gathers
summer flowers

and then
puts them in my hair.

We walk on back with shoes in hand
'cause in the summer heat,
the cool, green grass, we both agree,
feels good on our bare feet.

And as we near the house again,
we see someone has set
the table on the patio
and, no, he didn't forget
to set an extra place for me.
Oh! What a nice surprise!

Grandpa's made
a feast for me.

❖

Hmm!
Grilled burgers with french fries.

❖

The three of us, we dine outside
and while we're eating slow . . .

. . . we take the time
to share our thoughts

❖

before I have to go.

We hear my mother's car drive in,
that means I must depart.
My Grandma whispers, "Thank You,"
as she holds me to her heart.

She waves good-bye as we drive out.
Then from the road, I hear,
"You are an angel," Grandma shouts.
I hear her loud and clear.

I smile, wave back. I think it's true.
What Grandma said is right.

I know the light within me now
will shine forever bright.

Grandma's Oatmeal Cookies

Ingredients:

3 cups quick oats, uncooked
1 cup all-purpose flour
1/2 teaspoon baking soda
1 teaspoon baking powder
1/2 teaspoon salt
1/4 teaspoon cinnamon

3/4 cup butter flavored shortening
1 1/4 cups light brown sugar
1 egg
1/3 cup milk
1/2 tablespoon vanilla
1/2 cup raisins

Baking Instructions:

1. Turn music on or pick favorite song to sing.
2. Together, (or taking turns) Grandma and grandchild sing into spoon. (A little dancing and clapping provides added flavor.)
3. Position child on chair for easy mixing.
4. Grease baking sheet with shortening or use cooking spray.
5. Combine butter shortening and light brown sugar in large bowl. Cream together with singing spoon.
6. Add egg, milk and vanilla to creamed mixture. (Remember, a few bits of egg shells accidentally dropped in batter provide for added surprising crunch)
7. Combine oats, flour, baking soda, salt, baking powder and cinnamon. Mix into creamed mixture.
8. Add a 1/2 cup raisins (sneak a few)
9. Drop by rounded tablespoons 2 inches apart onto baking sheet.
10. Bake one baking sheet at a time for 10 to 12 minutes or until lightly browned at 375°.
11. Leave cookies on baking sheet for 2 minutes then put them on a sheet of foil to cool.

Makes 3-4 dozen depending on scoop size

For extra rich taste and enjoyment, be sure to share old stories while cookies bake.
Best when served on back patio or deck (weather permitting) with a glass of milk and Grandpa.